Powerships, Power Barges, Floating Wind Farms -
Electricity when and where you need it.

Patrick H. Stakem

(c) 2021

Table of Contents

Introduction..3
Author...4
The generation of Electricity..5
The transmission of Electricity...8
Powerships..11
Power Barges..15
Site Preparation and connection to the Customer's Grid.........17
Floating Wind turbines...17
Who are the players?..18
Floating offshore windfarms..20
Power sources, nuclear...22
Nuclear Power Ships – a good idea?.......................................22
The commercial world..24
Afterword..24
Glossary..25
Bibliography...28
References...31
Resources..32
Glossary of Nautical Terms and Phrases.................................33
If you enjoyed this book, here are some other from the author....41

Introduction

The book is about an energy technology solution called Powerships, and some of the alternate solutions they provide. There is a healthy growing industry in this field, with established and trusted suppliers. More than 60 such ships have been deployed around the world so far. They don't provide a long term solution, but are ideal in the short term, to assist in addressing peak loads, and to support infrastructure while ground-based plants are built. In natural disasters, they can support the grid while extensive repairs are made. Before we get into the ships, lets take a look at electricity, and the generation and consumption of it.

The first electrical systems used the direct current format of Nikola Tesla. There were problems with distance, due to excessive loss in the copper wire. It would work in a factory, but not in a city block. Direct current is also a form of flowing energy that can be captured in a battery.

A superior system was used by Thomas Edison, the alternating current scheme. Here, the source, called the alternator, was simpler than its dc generator counterpart. He used a frequency of 60 Hertz. In the alternator, the voltage is limited by the strength of the insulation on the wires. You want to make the voltage as high as possible for transmission, to reduce losses. The advantage of alternating current is that you can use a transformer to do this, and another transformer at the load end to bring the voltage back down. You've seen these cylindrical transformers up on power poles.

Edison's first power plant used six dynamos and could produce 100 kilowatts, which could service a square mile of New York real estate. Edison also invented the electric meter, so he could bill his customers based on usage.

Author

Mr. Patrick H. Stakem has a degree in Electrical Engineering from the Carnegie Mellon University, and Masters degrees fin Applied Physics and Computer Science from the Johns Hopkins University. He spent 43 years working on various projects at all of the NASA Sites.

During his undergraduate years, he worked summers for the local electrical company, in the substations group. The company, at the time, was replacing groups of three single-phase transformers with a newer three-phase transformer. Here, he learned the hands-on of coal-fired generation, voltage step-up transformers for transmission, and "reclosers" which acted like circuit breakers at high current and voltage. In a hands-on moment, his crew chief sent him up an aluminum ladder in an active substation. Not even to the top, the fluorescent tube was growing brightly. Electric fields! They exist.

Mr. Stakem can be found on Facebook and LinkedIn.

Comments, corrections, suggestions are appreciated.

The generation of Electricity

Generally, electricity is used when it is generated. We run into the scenario that we need power at night for lighting, but our solar arrays are not generating. We have to be clever, and store energy when it is available, for use when it is needed.

No storage/source will be 100% efficient, so we loose some of our daytime power as we draw it out at night.

In the simplest case, during the day, we use solar generated power, and charge a battery with what we don't need. At night, we operate from the battery, using an inverter to get back to AC. We must size the system so the heaviest load at night will operated properly in the hours of darkness, and that the solar panels have enough power to charge the battery's on a cloudy day.

Wind turbines generate AC power, usually 60 Hz and 600 volts. A transformer is used to supply the appropriate voltage to the grid. The turbines have a minimum "cut-in" speed, where they can start providing useful power This is around 10 MPH.

Similarly, there is a certain minimum amount of brightness for a solar panel to supply useful energy.

Today, the electrical grid is alternating current (AC).

That was not always case. The very early market and technology was direct current, the favorite of inventor Thomas Edison. He was the expert, and defined the industry and the components. He got the first successful light bulb lit up in 1879. Now he could sell these to businesses and households, and then sell them the electricity to make them work. In 1880, he formed a company, the Edison Illuminating Company. His first electric utility was the Pearl Street Station in New York City. He supplied 110 volts of direct current to some 59 customers in Manhattan. In 1882, he opened a steam-power generating station in London.

If you didn't like DC, invent your own and capitalize it. As a proponent of alternating current and one-time Edison employee, Nikoli Tesla did just that.

Tesla was a Serbian American who contributed greatly to the development and deployment of AC power. He became a naturalized citizen of the U. S. in 1884. He worked for Edison in New York City briefly. He later worked on wireless transmission of power, and communications.

Other early inventors such as George Westinghouse were also convinced that AC was the way of the future. In what has been called the Current Wars, both sides pointed out how dangerous and inefficient the other side's system was. Tesla's design for an poly-phase AC motor was licensed by

Westinghouse. Edison's DC system only served customers within 1 mile of the generating plant. Despite the drawbacks, Edison's marketing brought lighting, refrigerators, washing machines and such to small community's across America (including the Author's home town, Cumberland, Maryland).

In what came to be called the War of the Currents, the established DC systems faced off with the upstart AC systems. Transformers were built by Westinghouse, to boost the voltage for transmission, and lower it at the customer's premises for safe use. Thinner wires were needed (although better insulation) for AC. In 1887, Westinghouse had 68 AC stations, to Edison's 121 DC stations. By 1892, the war of the currents came to an end. Edison was bought out, and a new company, General Electric, was formed.

But, alternating current was ultimately the right answer, and DC disappeared except for some trolley systems. Locomotives used dc until switching systems that could handle the high currents involved. All modern locomotives use AC power, and some are hybrid systems with a battery and inverter.

Why did ac win? Early systems were dc, because it was easier. The problem was, there is no way to distribute it widely. You might have a coal fired system for a small town, but each town needed one.

It's also hard to switch large currents. They tend to melt the switch. For AC, you switch when the current goes to zero, momentarily. Out of high school, my father worked for Westinghouse, in the design of 'mercury arc" rectifiers. These could quench the big arc when switching large currents. They were used in the conversion of ac to dc.

Generally, all real loads have resistance, capacitance, and inductance. If the current is in phase with the voltage, you have real (can do work) power. Since the load has capacitance and inductance, the current will be out of phase with the voltage. This produces what is called reactive power, which does no work. It just flows between the source and the load.

An AC system generates alternating current. If this is fed to a strictly resistive load (like a space heater), the generated power is transformed into heat. If the load is not purely resistance, it will have the property's of capacitance and inductance as well.

The transmission of Electricity

The electric transmission and distribution system is called the grid, due to its architecture. They are individual switches and safety mechanisms, used to minimize damage in the case of lightning strikes or downed wires. You've seen the big tall towers carrying the transmission wires. The top-most wires are grounded, making them the target for lightning. The transmission is two-phase, using 2 wires with a neutral, 3 wires required. The frequency is 50 or 60 Hertz, depending on where you're at. The voltage is 10's of thousands of volts. Raising the voltage minimizes the line loss, which is due to the resistance of the wires. Transformers at the generating plant raise the voltage for transmission, and at a substation, which might serve a community, other transformers reduce the voltage for local distribution.

For example, I live in the United States, so my frequency is 60 Hertz, handy for running electric clocks. The wires that enter my house are two-phase, rated for 100 amps. There are three wires, a neutral, and two power wires. These come into an electrical box, where there is a bank of circuit breakers. Generally, large consumers of electricity such as air conditioning, stove, dryer, get their own circuit breakers. Other breakers may be for a room's outlets and lights.

A really good idea is to have a "whole house" surge suppressor installed at the box. If lighting hits the lines, that unit dies to protect all the house wiring and appliances. Otherwise, your TV and computer would do it.

The grid is managed across the country. In the U. S., there are three grids, East, West, and Texas. Texas is a historically independent state.

From a central control, the grid is monitored and controlled, and can be reconfigured if some portion of it has failed.

Rarely do you have or even want a generating plant next to your home or businesses. Particularly nuclear (accidents do happen) or coal burning. This means we need an infrastructure of transmission equipment. The big problem is, the voltage needs to be a high level (100,000's of volts), to keep the transmission losses low. Power is voltage times current. With high voltage, we have a correspondingly lower current. Resistive losses in the wires is related to resistance squared. As much as the early dc power systems enabled things like lighting not involving gas, washers and dryers, and refrigerators, you had to be within a few miles of the generating plant. You could use a gas-refrigerator, my Aunt had one, but it still needed to be plugged into the wall outlet, to run the compressor.

Electrical power transmission is inefficient for low-voltage, high current systems., For DC, you can't easily transform the power into higher or lower voltage, it has to be generated at those points. In physics, Joule's first law says the heat generated by an electrical conductor is proportional to the product of the resistance times the current squared. Thus, we want to keep the current low, and, to achieve the same power, make the voltage high. This is difficult to do with direct current.

With AC, we can readily take what we generate, boost the voltage, lower the current, and reduce the voltage back to a standard (110 volts for the US) for safe usage in user devices. The first transformer was in use in 1881.

Our power system, or grid, consists of sources of power, a distribution network, and users of power. If there is no storage, then electricity is used as it is generated, and the system must respond to varying loads. By 1836, the telegraph system used battery's. Pumped storage hydro was used by Connecticut Light & Power in 1929, a 31 megawatt pumped storage system. Power was also stored in flywheels. Today, pumped hydro represents the majority of power storage around the globe.

Direct current can be stored in battery's, but AC has to be stored in potential energy systems such as pumping water back uphill to the hydro dam. There is no free lunch. Each method of generation and storage is slightly inefficient, and some power is lost. We can convert ac to dc or vice versa with a motor-generator set, where, for example, a DC motor would drive an AC alternator. AC can be turned into DC by the rectification process. In the early trolley era, the motors were dc, and commercial AC power was used, rectified at the trolley facility. Most trolley systems today use a new generation of efficient AC motors. Locomotives, for example, use onboard diesel engines to turn alternators which supply the motors at each axle. A feature called "dynamic braking" is used, when going downhill, to control speed. Essentially, the motors are driven in reverse, and generate power which is dissipated in large banks of resistor grids on the roof.

Powerships

Developing Nations and Nations with an existing but insufficient electrical grid infrastructure can use power ships. These can provide short term solutions while existing infrastructure is repaired or replaced.

At this writing, there are 60 or more floating power stations operating in various locations on the planet. Most are privately owned, and leased out for profit. They are designed to operate continuously at full load. Originally built in older freighters, they are now using hulls specifically designed for their mission. Using an existing hull, a power ship can be outfitted in about 8 months.

Perhaps the first Powership was the SS Jacona, called at the time a floating electric power plant.

Jacona started out its maritime career as a 7,000 ton steam-powered cargo ship built in 1919, by the Todd Dry Dock and Shipbuilding Co. of Tacoma, WA, for the U. S. Shipping Board. It was rebuilt in 1931 by Newport News Shipbuilding in Virginia. The customer was the New England Public Service Company in Augusta, Maine. Tired of severe winter storms taking out transmission plans, the president of the Public Service Service System came up with the radical idea. It also provided power to popular seaside resorts in the summer months.

The idea for a floating power plant was from Walter Scott Wyman with the engineering design by the North Eastern Energetic Process Services Co. (Nepsco). Wyman was one of the founders of Central Maine Power, and studied Electrical Engineering at Tufts University. Interestingly, Wyman didn't graduate, but left to formed Central Maine Power, which produced electricity by the water from the Wyman dam.

The ship had dual 10,000 kilowatt steam turbines. She went to Bucksport, Maine, to supply electrical power to a paper mill. This arrangement was for less than a year, until the mill could be tied into the grid.

The Jacona could steam close to an effected area, moor, and hook into the local grid.

The U. S. Army had converted a Liberty Ship into a floating nuclear power station, called MH-1A Sturgis. It was to be used to supply electricity remote, nearly inaccessible sites. It went operational in 1967, and was towed to the Panama Canal Zone. It operated from 1968 to 1975, and was decommissioned.

The Liberty Ships' engines were removed, and the ship was towed to a semi-permanent location. The reactor was a pressurized water reactor using enriched uranium, inside of a 350 ton containment vessel. It was built by Martin Marietta, starting 1n 1963. The reactor was built by the American Locomotive Company (Alco). They had previously built a reactor for the Fort Belvoir in April 1957.

The Army was evaluating units that could supply heat and electrical power to remote locations. Early checkout of the system took place at Gunston Cove, in Fort Belvoir, Virginia. It spent 11 months supplying power to Fort Belvoir during checkout. The ship went to the Panama Canal Zone and supplied 10 megawatts of power to the Canal Zone from 1968 to 1975. It was particularly useful during a water shortage in 1968. It stood in for the Hydropower lost due to low water levels.

The reactor was refueled after 1 year of operation. It headed back to the U.S. At then end of 1976, but needed to stop along the way at Military Ocean Terminal Sunny Point, NC for structural repairs. By March 1977, it was at Fort Belvoir, where the fuel was removed from the reactor, and sent to the Savannah river site. The ship was then moored at Fort Eustis, Virginia as part of the James River Reserve Fleet. She was decommissioned in 2014, It then went to Galveston, Texas, for removal of radioactive scrap. She then made her last voyage to Brownsville, TX, for scrapping.

The complications of nuclear reactors, have held back their use in commercial power plants, although there has been some recent interest by Russia and China.

Powerships are not intended as a permanent solution, but rather to handle peak demands, and fill in the gap while shore based systems are built or repaired.

By the end of 2010, Pakistan was benefiting from a Turkish Powership, the Kaya Bey, capable of 230 MW. An older unit, it burns furnace-grade oil. It is anchored off of Karachi.

There is a five year contract with the heavily indebted National Power Company. The ship will only make a dent in the country's power crisis, not solve the problem. A total estimated additional 5,000 megawatts is needed. Pakistan relies heavily on hydro-power, but doesn't have the money to invest in the construction of new plants.

The USS Saranac was a fleet oiler during World War-2. After the war, she was sold as surplus, and became a floating power generation facility stationed at Kobe, Japan.

South Africa is another country benefiting from the use of commercial Powerships.

Power Barges

There are currently some 75 Power Barges operating around the world. The trend is toward Power Barges over Power Ships due to the reduced cost – no engines required. The space taken up by the ships engines could host another generator or two. However, there is the issue of having to hire a tug to move the ship.

Generally, these do not include crew quarters, as the crews are brought out to the barge as needed. The equipment continuously monitors itself, and can send out a message if there is a problem.

It may surprise you, particularly if you live in Brooklyn, N.Y., that a 640 megawatt generating station is sitting on 4 power barges in Gowanus Bay. There are two additional barges at Sunset Bay. It is one of the largest such facility's in the world. The four barges are each fitted with 8 turbines. They tie into a ConEd substation with a 138 KV line. The facility was one of the first to be back in operation after hurricane Sandy devastated New York City. The turbines can use natural gas or low-sulfur diesel. They keep about 1.5 million Brooklyn-ites happy and using their appliances. If the units are shut down, they can be providing power within 10 minutes.

Not having engines, the power barges need to be towed into position, and moored to a pier. There are certain advantages. Power ships don't use their engines after they reach their destination, and are moored in place. They are more expensive to built. Power Barges are more cost-effective overall.

New York City is getting two new power barges that will use the latest gas turbine technology. Each ship will be outfitted with eight Siemens 76 Megawatt gas turbines, derived from aviation.

Specifically, the turbines are a variation of a Rolls Royce aircraft engine, used on the Boeing 777. These barges are called *SeaFloat* by Siemens. They engines operate at 41.8% efficiency, using natural gas. They can switch from gas to liquid fuel and vice versa while continuing to operate. They supply around 60 megawatts. Floating power plants utilizing Westinghouse gas turbines date back to the 1990's. The barges come with a 20-year service agreement.

Siemens is also working on a 145 mWatt power barge for the Dominican Republic. This will include a Fluence Energy battery storage system. These are generally install to respond to peak energy demand. These can supply from 2 to 500 megawatts for 1-6 hours. They can supply either 50 or 60 Hertz AC with the built-in inverters, for 6 hours. They use lithium ion battery's, like Tesla's Powerwall. The Fluence devices have their own built-in software for control, management, fault detection.

The concept of floating power plants is also viable for servicing areas with no existing electrical service.

Construction of Power Ships and barges is a thriving business. Company's such as Power Barge Corporation, BWSC, Hyundau, and IHI Corporation work in the field of construction of the facility's, as does MAN, Wartsila, and Karadeniz Energy. Most units now are gas turbine powered, sometime with the dual fuel feature.

Site Preparation and connection to the Customer's Grid.

The ship will require a dock to tie-off to when it arrives, or it may be secured a short distance offshore, using multiple blocks on the seafloor. It will then have to bring a high tension cable ashore to connect into the customer's grid. If the ship is to be refueled from shore via a hose, that needs to be properly secured. If the ship is not moored directly alongside a pier, the crew will need to use a light boat to come onshore, and to resupply provisions. The ship can also be re-provisioned from a light boat. This would be used in a pandemic situation, where the power boat crew could be isolated.

The ship may also serve as the source of potable water for the client, with onboard desalination equipment.

Floating Wind turbines

A Ship can also be outfitted with wind turbines and

towed to a favorable location. Fixed wind turbines are in use off the coast of many country's. The technology is mature. Wind Farm vessels are emerging as yet another solution to growing energy needs.

Who are the players?

Here are some of the big players in the Power Ship & Power Barge world.

A private company in Istanbul, Turkey, Karpowership has the largest Powership. fleet They have provided more that seventeen Powerships with a total capacity of 2.8 G watts. For them, everything is done in-house. Then tend to buy older freighters, and convert them, based on an in-house architecture and design Group.

Their first vessel was acquired in 2007, a freighter turned into a floating power plant at the Sedef shipyard in Istanbul, It flys a Liberian flag, and uses 12 diesel engines to turn the generators.

Karpowership is a division of the Karadeniz Energy Group of Istanbul, Turkey. It is a large operation, with a vast fleet of 25 Powerships available to customers with a combined total capacity of more than 4 gigawatts. They are currently working on doubling the size of their fleet. Since 2010, they completed fifteen new ships, with a combined capacity of 2,800 megawatts. These were generally freighters, converted at a shipyard in Turkey.

They have ships serving around Africa, the Mediterranean, Indonesia, and Iraq. They are trying to make their operations plug-and-play. Generally, a

Powership will arrive within 120 days after a contract has been signed.

Waller Marine, of Houston, Texas is a major player in design and construction of floating power plants. It includes a companion 360,000 barrel floating diesel storage Barge. They have the distinction of producing the world's largest floating power plant, a 342 MW unit in Venezuela.

Most of their units use industrial gas turbines, such as the General Electric 7FA. They are working on a 1,000 Mwatt design.

BWSC (Burmeister & Wain Scandinavian of Denmark), focuses on fuel efficiency. Their onboard liquid fuel tanks hold enough fuel for 2.5 days of operation. They use radiators instead of sea water for cooling. Capacity's range from 60-150 KW. Waste heat recovery boilers are used for fuel heating. The barges have a draft of 3.6 meters. Turbocharged diesel engines are used. The barge's systems feature built-in diagnostics. They have 180 power plants in 54 country's.

Power Barge Corporation is focusing on fast track, plug-and-play systems, based on standardization. The company was established in 2000 to address the growing field of floating power plants. Their turbine-based units can use natural gas, diesel, bio-diesel, or fuel oil. Power generation ranges to 200 Mwatts.

They also are addressing the concept of Energy Storage Power Barges. These use high efficiency Lithium-Ion technology, and can hold up to 1,500 Megawatt-hours of energy.

MAN Energy Solutions is a multinational company located in Germany. It is a producer of large diesel engines and turbochargers for marine use. It has consolidated with the Danish Company Burmeister and Wain, shipbuilders. Their diesel engines range up to 87 MW. They also offer gas turbines of up to 50 MW capacity. These are used to provide entire electrical power plants.

General Electric is a major source in the wind turbine arena. The Haliade-X is specifically designed for offshore use. It comes in 12, 13, or 14 Megawatt versions, with 107 meter blades. GE Renewable Energy maintains its production site for offshore wind turbines in Saint-Nazaire, France.

Vestas Wind Systems A/S is a Danish Company in the wind turbine field since 19479. It is now the largest wind turbine company in the world. It has installed over 66,000 units with a combined capacity of 100 Gwatts. The Vestas V164 is the most powerful wind turbine at 9 megawatts capacity.

There are some 39 current manufactures of wind turbines, across the planet, with new ones entering the field.

Floating offshore windfarms

Offshore wind energy capture is entering a new phase. In addition to stationary wind turbines anchored to the sea floor. These can be used in deeper water than those built on platforms. The goal is investing $70 billion to provide new capacity of 18.6 gigawatts by 2030. Two emerging company's, Diamond Offshore Wind and RWE Renewables, have joined in a partnership with the State of Maine to start the process. Diamond is a subsidiary of Mitsubishi, and RWE is a German utility. A new American company, Principle Power Inc. did two installations in Portugal, and another off the coast of Scotland. There is even a new trade group, European Ocean Energy Association. There is a proposal to build a floating wind farm some 30 miles off the California coast, in the central part of the state. The platforms will be of the U. C. Berkeley Windfloat architecture. These use water pumped between ballast tanks to keep the platform upright at all times.

Hywind Scotland is the world's first commercial wind farm, located 29 km off the coast. Each facility has five of the 6 megawatt turbines for a total capacity of 30 MW. The turbines are from Norway. They are anchored to the seafloor with suction anchors. These are a fixed platform anchor with an open bottom tube embedded in sentiment, sealed at the top. Lifting forces result in a pressure differential which holds the anchor down. These are widely used in the offshore oil and gas industry. The turbine is mounted on a tension-leg platform. These use tethers anchored to the seafloor, and resist any vertical movement of the platform.

The first operational deep-water floating wind turbine was deployed in 2009. It had a 390 foot tower supporting a 2.3 Mwatt generator. The platform was 6.2

miles offshore.

The first floating offshore wind farm in the Mediterranean is to be located in the strait of Sicily. This 250 Mwatt facility will have 25 turbines of 10 Mwatts capacity each. The cost of the project is around 740 Euros. The facility was built in Denmark, and is situated to not be visible from the shore.

Power sources, nuclear

Power ships can use nuclear reactors, engines supplied with a hydrocarbon gas such as methane, or petroleum based fuels. These are consumed in the engine to get the power to spin large turbines to generate the electrical power. The ship has storage facility's for the fuel, or could use a pipeline from the shore., or from a tanker ship. Another fuel source is petcoke, a "waste" material from petroleum production. Some engines can operate on dual fuels.

Nuclear Power Ships – a good idea?

The United States built the first nuclear power station in a ship in the late 1960's. It could product 10 megawatts of power.

As long as you have control over a nuclear power plant, the bad guys can't steal your fissionable material and build nuclear bombs. This works as long as you have physical control over the plant. This is much easier if the plant is fixed, and in your own country. There is major security risk with nuclear plants, both due to terrorists, and, as the Japanese found out, Tsunami's.

Russia has a ship-based nuclear power plant in the far Arctic at the east end of their country. The Akademik Lomonosov has two 150 megawatt reactors. It was delivered 8 years behind schedule, and was a factor of three over budget.

They had an ambitious plan to build 20 floating nuclear power plants for the South China Sea region. Floating nuclear plants have the advantage of access to large volumes of cooling water for the reactors. The water is desalinized first to prevent corrosion in the cooling tubes. It can then be cooled, and supplied as potable water.

China constructed its first floating roject in 2016, with a plan to support China's control over certain regions in that area.

The Russians are building floating nuclear power plants, with an eye to mass production. The first unit, the Akademik Lomonosoc was operational in December of 2019 in the Russian Arctic. It was built in a submarine building plant, and later transferred to the Baltic Shipyard. If it works out, seven more are planned. It was launched in 2010. The floating power station does not have its own propulsion. It carry's a crew of 69, and has two reactors based on naval reactors for Ice Breakers. It can provide 70 Megawatts, or 300 megawatts of heat for a city. It can also serve as a desalination plant.

The reactor needs to be refueled every three years, but has a service life of 40 years.

The commercial world

There are many commercial entity's around the world building and operating Powerships successfully. There is no doubt this is a growth industry, and the demand for electricity continues to rise, Power Ships and wind farms use the latest technology to achieve even greater efficiency and decrease pollution compared to contemporary facilitys.

Afterword

The future of Powerships is definitely Power Barges, with no permanent crew or their facilities. Onboard watchdog software will monitor systems, and call for help when needed. One can argue there would be a quicker response to situation, if there was at least one crew member onboard at all times.

Power Barges are rapidly improving, taking advantage of new turbine technology to replace aging internal combustion engines.

Floating wind farms are growing rapidly as an option for electrical needs. There are products of combustion released, and the platforms can monitor themselves. Placed out much further from shore, these facility's can supply needed energy without being in sight of land. We can also utilize wave motion as a power source to generate electricity.

It should be interesting. Hang on tightly.

Glossary

AC – alternation current.
Active power - "real power" measured in watts.
AEP – Annual energy production
Aeroderivative – derived from aircraft technology.
Ampere – a unit of electrical current flow.
Apparent Power – product of the rms value of voltage and current. Units of volt-amps.
Arc – electrical discharge through an ionized gas.
ASIN – Amazon Standard Inventory Number.
BESS – Battery Energy Storage Systems.
Black start – ability of a facility to start without external power.
BOEM – (U. S.) Bureau of Ocean Energy Management.
B&W – Burmeister & Wain Shipbuilders
ConEd – Consolidated Edison Co. New York.
COP – Copenhagen Offshore Partners
DC – direct current.
DOE – (U.S.) Department of Energy.
Dual fuel engine –can use liquid or gas fuels
Dynamo – a machine to convert mechanical energy into electrical energy.
EPC – Engineering, Procurement, and Construction.
Floater – floating offshore wind turbine
FPP- floating power plant.
FPSO – Production Storage and Offloading Units.
IGCC – Integrated Gasification Combined Cycle
IPP – Independent Power Producer.
Kilowatt - 1,000 watts.
Gas turbine – example is a jet engine for aircraft

Giga watt 10^9 watts

Giga watt hour – measurement of energy.

HPO – hydrogenated palm oil

HRSG – Heat Recovery Steam Generator.

LNG – liquefied natural gas.

LS – low sulfur.

Load shedding – the process of cutting off some loads so that others can be addressed.

Megawatt – 1,000,000 watts.

MG – marine grade.

NOAA – (U. S.) National Oceanographic and Atmospheric Administration.

NPP – nuclear power plant

NREL – National Renewable Energy Laboratory (U.S.)

OCGT – open cycle gas turbine.

Peaker – electrical plant to handle peak power demands.

Petcoke – a solid material derived from oil refining.

Power Barge – A power ship with no engines of its own. It is owed into position and anchored.

Power factor – ratio of actual electric power in an AC circuit to the product of the rms values of the voltage and current. Difference is due to the reactance in the circuit.

PPA – power purchase agreement

Reactance – opposition to the flow of current due to inductance or capacitance.

Reactive power – power which flows back and forth from the load, but does not do work. Units of var.

RMS – root mean square

Spinning reserve – generation capacity that is online, but not connected.

TLP – tension leg platform
Transformer – a device to step up or down a voltage.
VAR – volts-ampere reactive.
VAWT – vertical axis wind turbine
Volt – a unit of electrical potential
Watt- a unit of electrical power. Voltage times current.

Bibliography

Andrea, Davide *Battery Management Systems for Large Lithium Ion Battery Packs, ISBN*-978-1608071043.

Asian Development Bank, *Handbook on Battery Energy Storage System*, 2018, ISBN-978-9292614706.

Blume, Steven W. *Electric Power System Basics for the Nonelectrical Professional,* 2016, ISBN-978-1119180197.

Breeze, Paul *Power Generation Technologys*, 3rd ed, ISBN-978-0081026311.

Crawley, Gerald M. *Energy Storage (World Scientific Series in Current Energy Issues),* 2017, ISBN-978-9813208957.

Glover, J. Duncan; Overbye, Thomas; Sarma, Mulukutla, S. Sarma *Power Systems Analysis and Design*, 6th edition, ISBN-13: 978-1305636187.

Gonen, Turan *Electric Power Distribution Engineering*, 3rd Edition, ISBN-13: 978-1482207002.

Henami, Ahmad *Wind Turbine Technology*, ISBN-978-1435486461.

Jones, Lawrence E. *Renewable Energy Integration: Practical Management of Variability, Uncertainty, and*

Flexibility in Power Grids, 2017, ISBN-978-0128095928.

Kirschen, Daniel S. ; Strbac, Goran *Fundamentals of Power System Economics 2nd Edition*, ISBN-978-1119213246.

Kundur, Prabha *Power System Stability and Control,* 1st Edition, ISBN-978-007035958.

Matt, Stephen R. *Electricity and Basic Electronics*, 8th ed, ISBN-978-1605259536.

Miller, Robert Malinowski, James; *Power System Operation,* 3rd ed, ISBN-13 978-0070419773.

Ng, Chong; Ran, Li (ed) *Offshore Wind Farms: Technologies, Design and Operation* (Woodhead Publishing Series in Energy), ISBN-978-0081007792.

Rogers, E. G.. (1931). S.S. Jacona. Southern Power Journal. W. R. C. Smith Publishing Company. p. 50. "The Jacona, the first successful conversion of a ship into a floating power plant, has been recently placed in service by the New England Public Service Company of Augusta, Maine."

Stoft, Steven *Power System Economics: Designing Markets for Electricity*, 1st Edition, ISBN-13: 978-0471150404.

Svenvold, Mark "The world's first floating wind turbine goes on line in Norway". DailyFinance.com.9 September 9, 2009

Tacx, Jochem *Building an Offshore Wind Farm: Operational Master Guide - Limited Edition,* 2019, ISBN-1710288825.

Thomas, Mini S.; McDonald, John Douglas *Power System SCADA and Smart Grids,* 1st Edition, ISBN-13: 978-0367658847.

Thomsen, Kurt *Offshore Wind: A Comprehensive Guide to Successful Offshore Wind Farm Installation,* 2nd Edition, ISBN-978-0124104228.

U. S. Department of Energy, Office of Scientific and Technical Information, "Feasibility of Floating Platform Systems for Wind Turbines," avail: https://arc.aiaa.org/doi/10.2514/6.2004-1007

Wood, Allen J. ,Wollenberg, Bruce F., Sheble, Gerald B. *Power Generation, Operation, and Control,* 3rd Edition, ISBN-978-0471790556.

Wu, Fu-Bao (ed) , Yand, Bo (ed), Ye, Ji-Lei (ed) *Grid-Scale Energy Storage Systems and Applications*, 1st Edition, ISBN-978-0128152928.

References

American Bureau of Shipping, *Guide for Building and Classing Floating Offshore Wind turbine Installations*, Updated, July 2014.

Avail: https://ww2.eagle.org/content/dam/eagle/rules-and-guides/archives/offshore/195_floatingoffshorewindturbineinstallations/FOWTI_Guide_e-July14.pdf

Hearst Magazines, (February 1931), "A Floating Power Plant," Popular Mechanics. pp. 217–218.

Marine Engineering (1930). "Floating Power House Jacona." Marine Engineering & Shipping Age, n35. Aldrich Publishing Company. p.406.

Viselli, Anthony; Dagher, Dr. Habib; Dwyer, Mark *Floating Wind Turbine Platform and Method of Assembling*, 2013. avail: https://composites.umaine.edu/publication/floating-wind-turbine-platform-and-method-of-assembling/

U. S. Atomic Energy Commission, *A survey of unique technical features of the floating nuclear power plant concept*, 1974, ASIN – B0030T1C9G.

U. S. Government Accountability Office, *Before Licensing Floating Nuclear Power Plants, Many Answers Are Needed, 2019,* ASIN – B07V3YWCWM.

U. S. Nuclear Regulatory Commission, F*inal*

environmental statement related to the proposed manufacture of floating nuclear power plants, Offshore Power Systems, Docket no. STN 50-437, 1978, ASIN-B002ZVP99E.

Resources

Global Powership Market:

https://www.reportlinker.com/p05791394/Global-Powerships-Market.html?utm_source=PRN

www.repoweringbrooklyn.com

Wikipedia, various.

Glossary of Nautical Terms and Phrases

ABAFT - behind, or farther aft. The mainmast is abaft the foremast.

ABEAM - at right angle s to the center line of the ship. Refers to an object outside the ship.

ABOARD - on or in a ship.

ACCOMMODATION LADDER - steps leading down a ship's side ; used for boarding.

ACKNOW LE DGEMENT - a statement that a message has been received.

AFT: at, near, or in the direction of the stern.

ALL HANDS - refers to every man (person) aboard .

AMIDSHIPS - in the line of the keel;sometimes halfway between the bow and stern.

ASTERN - behind the ship ; on a bearing of 1800 from ahead.

AUXILIARY - an assisting machine or vessel, such as an air-conditioning machine or a fuel ship.

BALLAST - heavy weights in the hold of a vessel or ship to increase stability by lowering center of gravity.

BEACON - an aid to navigation placed on or near a danger spot.

BEAM: the greatest width of a ship.

BEARING - the direction of an object expressed either in degrees either as relative or true bearing.

BELOW - beneath the main deck.

BERTH - a space for a ship to moor or anchor.

BILGE - lower part of a vessel where waste and seepage collect.

BILLET - an allotted place to sleep ; refers also to a particular man's duties aboard ship.

BLINKER - a set of lights at the masthead or on the end of a yardarm, connected to a telegraph key and used for sending flashing light signals .

BOAT: a small vessel which can be hoisted onto or carried by a ship.

BO'SN'S CHAIR - a seat, consisting of a short board fastened in the end of a line, on which a man may be suspended for working aloft or over the side.

BO'SN'S PIPE: a small , shrill whistle used by the bo'sn's mate.

BOW - forward part of a vessel .

BRIDGE - the raised pl platform in the forward part of the ship from which the ship is steered or navigated.

BULKHEAD - a partition separating compartments ; corresponds to a wall in a building.

BUOY - a floating marker moored to the bottom which by shape and color conveys navigational information.

CABIN - the captain's quarters .

CAPSIZE - to overturn in a small boat.

CAST OFF - to let go .

CENTER LINE - imaginary straight line running from the bow to the stern of a ship.

CHART - a nautical map used as an aid in navigation.

CLEAR - to leave a port with all formalities concluded; to empty; to work clear, as of a shoal; to untangle.

CLOSE ABOARD - near to the ship.

COLORS - the national ensign.

COMMAND - term applied to a ship or ships under one officer; a directive indicating what to do and how to do it.

COMPANION WAY - passageway on board ship.

COMPARTMENT - a space below deck between bulkheads, corresponding to a room of a building.

COUNTRY - the space near to a compartment or quarters, such as wardroom country or officers' country.

COURSE - the direction steered by a vessel expressed in degrees.

COXSWAIN - the enlisted man in charge of a boat and usually serving as steersman.

DAMAGE CONTROL - maintenance of watertight integrity of the ship during battle or storms, including necessary repairs.

DAVIT - a curved metal spar fitting in a socket on the deck and projecting over the side of the ship for hoisting boats or handling weights.

DEAD AHEAD - directly ahead.

DEAD RECKONING - a navigator's estimate of the ship's position dependent upon course steered and distance run, independent of sights or bearings ; derived from "deduced reckoning."

DECK - corresponds to the floor of a building.

DOCK - a landing pier for boats or ships .

DOG - a type of bolt and nut used to secure watertight doors, hatch covers, or manhole covers.

DOLDDRUMS - belts on each side of the Equator in which, ordinarily, little or no wind blows.

DOUBLE BOTTOMS - watertight subdivisions of a

ship next to the keel and between the outer and inner bottoms.

DOWSE - to put out a light ; to cover with water.

DRAFT - the depth of water from the surface to the ship's keel ; a detail of men.

EBB TIDE - condition along the coast when the tide is going out.

EMBARK - to go on board ship.

EMERGENCY SPEED - all the speed of which a ship is capable.

ENSIGN - the national flag; a junior commissioned officer in the Navy.

EVEN KEEL - floating level.

FAIR TIDE - a tide running in the same direction as the ship.

FAIRWAY - an open channel.

FANTAIL - the part of the stern of the ship.

FAST - secure.

FATHOM - a unit of measurement equaling 6 feet.

FLANK SPEED - a certain prescribed increment above standard speed.

FLYING BRIDGE - a bridge on a ship having no supports and extending out from the control tower.

FORE AND AFT - running in the direction of the keel .

FORECASTLE - the upper deck forward of the foremast; a forward compartment where the crew lives.

FRAME - the ribs of a ship strengthening and supporting the plating.

FUNNEL - the smokestack of a ship.

GALLEY - the ship's kitchen.
GANGPLANK - a portable bridge used for boarding a ship from a dock
GIG - ship's boat used by the commanding officer.
GUNWHALE - the upper edge or rail of a ship or boat's side.
HAND - a member of the crew.
HANDY-BILLY - a small portable pump.
HATCH - an opening in the ship's deck
HEAD - a ship's toilet.
HELM - a tiller
HIGH SEAS - the entire ocean beyond the three-mile limit where no nation has military authority
LINE OFFICER – an officer who holds military authority in the chain of command.
HULL - framework of a vessel together
MAIN DECK - the highest complete deck
INBOARD: toward the center line of the extending from stem to stern and side to side.
ISLAND - superstructure containing conning tower, navigation bridge, communication platform, flying bridge, etc.
JETSOM - goods which sink when thrown overboard at sea.
JETTY - a landing wharf or pier, a dike at a river's mouth.
KEEL - the backbone of a ship running from stem to sternpost at the bottom.
KNOT - one nautical mile per hour.
LADDER - a metal, wooden, or rope stairway.
LAND FALL - first sighting of land at the end of a sea

voyage.

LASH - to tie or secure.

LEAVE - special permission to be absent from ship or station.

LEE - away from the direction of the wind.

LIFELINE - a line secured along the deck to lay hold of in heavy weather;

MOORING - securing a ship to a dock or buoy, or anchoring with two anchors.

NAUTICAL MILE - the length of a minute of a great circle of the earth or 6,080.20 feet.

OFFICER OF THE DECK - the officer in charge of the ship during each watch and on deck as the captain's representative.

OVERHEAD - equivalent to the ceiling in a building.

PASSAGEWAY - a corridor or hallway onboard ship.

PILOT - an expert who conducts ships in and out of harbors and in dangerous waters.

POOP DECK - the partial raised deck and after structure at the stern over the main deck of a vessel.

PORT - the left side of a ship facing forward

QUARTERDECK - the part of the upper

STARBOARD - the right side of a ship deck reserved for honors and cere- looking forward.

STATION BILL - a bill listing stations of the crew at emergency drills.

QUARTERS - living space ; all hands assembled at established stations for review.

REEF - a chain or ridge of racks, coral, or sand in shallow water.

RUNNING LIGHTS - lights required by law which are

carried by a ship underway.
SCREW - propeller.
SCUTTLEBUTT - a container of water for drinking purposes; a rumor.
SEA KEEPING - the ability of a ship to stay at sea for long periods of time.
SECOND DECK - a complete deck next below the m main deck.
SECURE - to m make fast; to tie ; order given on completion of a drill or exercise on board ship.
SHAKEDOWN CRUISE - cruise of a newly commissioned ship to test all machinery and train the crew.
SICKBAY - ship 's hospital or dispensary
SKIPPER - slang for the captain.
SMOKING LAMP - the expression, "the smoking lamp it lit," means permission is given to smoke.
SOUND-POWERED PHONE - a phone powered by voice ; standard battle phone.
SQUARE A WAY - to get things settled down or in order ; to complete a job .
Staff officer – an officer whose duties are special rather than military; doctor or chaplain.
Starboard – right side of the ship, looking forward.
Station bill –a bill listing stations of the crew at emergency drills
Stern – after part of the ship
Superstructure – equipment and fittings, except armament, extending above the hull.
Swab – a rope mop
Swell – heave of the sea.

TAFFRAIL - a rail at the stern of a ship.
TOPSIDE - above decks.
UNDER WAY - a vessel is said to be underway when she is not at anchor nor made fast to the shore, nor aground.
UNDERTOW - a seaward current near the bottom in heavy surf.
WAKE - the track or trail in water which a vessel leaves behind.
WARDROOM - officers' assembly and mess room aboard a ship.
WATCH - a post or period of duty.
WATCH, QUARTER, AND STATION BILL - a list or chart giving the watch duties, billet and emergency, or battle station of every man aboard ship.
WEATHER DECK - the portion of the main, forecastle, poop, and upper decks which is exposed to the elements.
WINDWARD - into the wind ; toward the direction from which the wind is blowing.
YARD - shipbuilding and repair depot for Navy vessels- as Boston Navy Yard.
YAW - to steer badly, zigzagging back and forth across the intended course.

If you enjoyed this book, here are some other from the author.

Stakem, Patrick H. *16-bit Microprocessors, History and Architecture*, 2013 PRRB Publishing, ISBN-1520210922.

Stakem, Patrick H. *4- and 8-bit Microprocessors, Architecture and History*, 2013, PRRB Publishing, ISBN-152021572X,

Stakem, Patrick H. *Apollo's Computers,* 2014, PRRB Publishing, ISBN-1520215800.

Stakem, Patrick H. *The Architecture and Applications of the ARM Microprocessors,* 2013, PRRB Publishing, ISBN-1520215843.

Stakem, Patrick H. *Earth Rovers: for Exploration and Environmental Monitoring,* 2014, PRRB Publishing, ISBN-152021586X.

Stakem, Patrick H. *Embedded Computer Systems, Volume 1, Introduction and Architecture*, 2013, PRRB Publishing, ISBN-1520215959.

Stakem, Patrick H. *The History of Spacecraft Computers from the V-2 to the Space Station*, 2013, PRRB Publishing, ISBN-1520216181.

Stakem, Patrick H. *Floating Point Computation*, 2013,

PRRB Publishing, ISBN-152021619X.

Stakem, Patrick H. *Architecture of Massively Parallel Microprocessor Systems*, 2011, PRRB Publishing, ISBN-1520250061.

Stakem, Patrick H. *Multicore Computer Architecture,* 2014, PRRB Publishing, ISBN-1520241372.

Stakem, Patrick H. *Personal Robots*, 2014, PRRB Publishing, ISBN-1520216254.

Stakem, Patrick H. *RISC Microprocessors, History and Overview,* 2013, PRRB Publishing, ISBN-1520216289.

Stakem, Patrick H. *Robots and Telerobots in Space Applications*, 2011, PRRB Publishing, ISBN-1520210361.

Stakem, Patrick H. *The Saturn Rocket and the Pegasus Missions, 1965,* 2013, PRRB Publishing, ISBN-1520209916.

Stakem, Patrick H. *Visiting the NASA Centers, and Locations of Historic Rockets & Spacecraft,* 2017, PRRB Publishing, ISBN-1549651205.

Stakem, Patrick H. *Microprocessors in Space*, 2011, PRRB Publishing, ISBN-1520216343.

Stakem, Patrick H. Computer *Virtualization and the*

Cloud, 2013, PRRB Publishing, ISBN-152021636X.

Stakem, Patrick H. *What's the Worst That Could Happen? Bad Assumptions, Ignorance, Failures and Screw-ups in Engineering Projects, 2014,* PRRB Publishing, ISBN-1520207166.

Stakem, Patrick H. *Computer Architecture & Programming of the Intel x86 Family, 2013,* PRRB Publishing, ISBN-1520263724.

Stakem, Patrick H. *The Hardware and Software Architecture of the Transputer,* 2011,PRRB Publishing, ISBN-152020681X.

Stakem, Patrick H. *Mainframes, Computing on Big Iron,* 2015, PRRB Publishing, ISBN- 1520216459.

Stakem, Patrick H. *Spacecraft Control Centers,* 2015, PRRB Publishing, ISBN-1520200617.

Stakem, Patrick H. *Embedded in Space,* 2015, PRRB Publishing, ISBN-1520215916.

Stakem, Patrick H. *A Practitioner's Guide to RISC Microprocessor Architecture*, Wiley-Interscience, 1996, ISBN-0471130184.

Stakem, Patrick H. *Cubesat Engineering,* PRRB Publishing, 2017, ISBN-1520754019.

Stakem, Patrick H. *Cubesat Operations*, PRRB Publishing, 2017, ISBN-152076717X.

Stakem, Patrick H. *Interplanetary Cubesats*, PRRB Publishing, 2017, ISBN-1520766173 .

Stakem, Patrick H. Cubesat Constellations, Clusters, and Swarms, Stakem, PRRB Publishing, 2017, ISBN-1520767544.

Stakem, Patrick H. *Graphics Processing Units, an overview*, 2017, PRRB Publishing, ISBN-1520879695.

Stakem, Patrick H. *Intel Embedded and the Arduino-101, 2017,* PRRB Publishing, ISBN-1520879296.

Stakem, Patrick H. *Orbital Debris, the problem and the mitigation*, 2018, PRRB Publishing, ISBN-*1980466483*.

Stakem, Patrick H. *Manufacturing in Space*, 2018, PRRB Publishing, ISBN-1977076041.

Stakem, Patrick H. *NASA's Ships and Planes*, 2018, PRRB Publishing, ISBN-1977076823.

Stakem, Patrick H. *Space Tourism*, 2018, PRRB Publishing, ISBN-1977073506.

Stakem, Patrick H. *STEM – Data Storage and Communications*, 2018, PRRB Publishing, ISBN-1977073115.

Stakem, Patrick H. *In-Space Robotic Repair and Servicing*, 2018, PRRB Publishing, ISBN-1980478236.

Stakem, Patrick H. *Introducing Weather in the pre-K to 12 Curricula, A Resource Guide for Educators*, 2017, PRRB Publishing, ISBN-1980638241.

Stakem, Patrick H. *Introducing Astronomy in the pre-K to 12 Curricula, A Resource Guide for Educators*, 2017, PRRB Publishing, ISBN-198104065X.

Also available in a Brazilian Portuguese edition, ISBN-1983106127.

Stakem, Patrick H. *Deep Space Gateways, the Moon and Beyond*, 2017, PRRB Publishing, ISBN-1973465701.

Stakem, Patrick H. *Exploration of the Gas Giants, Space Missions to Jupiter, Saturn, Uranus, and Neptune*, PRRB Publishing, 2018, ISBN-9781717814500.

Stakem, Patrick H. *Crewed Spacecraft*, 2017, PRRB Publishing, ISBN-1549992406.

Stakem, Patrick H. *Rocketplanes to Space*, 2017, PRRB Publishing, ISBN-1549992589.

Stakem, Patrick H. *Crewed Space Stations,* 2017, PRRB Publishing, ISBN-1549992228.

Stakem, Patrick H. *Enviro-bots for STEM: Using Robotics in the pre-K to 12 Curricula, A Resource Guide for Educators,* 2017, PRRB Publishing, ISBN-1549656619.

Stakem, Patrick H. *STEM-Sat, Using Cubesats in the pre-K to 12 Curricula, A Resource Guide for Educators*, 2017, ISBN-1549656376.

Stakem, Patrick H. *Lunar Orbital Platform-Gateway*, 2018, PRRB Publishing, ISBN-1980498628.

Stakem, Patrick H. *Embedded GPU's*, 2018, PRRB Publishing, ISBN- 1980476497.

Stakem, Patrick H. *Mobile Cloud Robotics*, 2018, PRRB Publishing, ISBN- 1980488088.

Stakem, Patrick H. *Extreme Environment Embedded Systems,* 2017, PRRB Publishing, ISBN-1520215967.

Stakem, Patrick H. *What's the Worst, Volume-2*, 2018, ISBN-1981005579.

Stakem, Patrick H., *Spaceports*, 2018, ISBN-1981022287.

Stakem, Patrick H., *Space Launch Vehicles*, 2018, ISBN-1983071773.

Stakem, Patrick H. *Mars*, 2018, ISBN-1983116902.

Stakem, Patrick H. *X-86, 40th Anniversary ed*, 2018, ISBN-1983189405.

Stakem, Patrick H. *Lunar Orbital Platform-Gateway*, 2018, PRRB Publishing, ISBN-1980498628.

Stakem, Patrick H. *Space Weather*, 2018, ISBN-1723904023.

Stakem, Patrick H. *STEM-Engineering Process*, 2017, ISBN-1983196517.

Stakem, Patrick H. *Space Telescopes,* 2018, PRRB Publishing, ISBN-1728728568.

Stakem, Patrick H. *Exoplanets*, 2018, PRRB Publishing, ISBN-9781731385055.

Stakem, Patrick H. *Planetary Defense*, 2018, PRRB Publishing, ISBN-9781731001207.

Patrick H. Stakem *Exploration of the Asteroid Belt*, 2018, PRRB Publishing, ISBN-1731049846.

Patrick H. Stakem *Terraforming*, 2018, PRRB Publishing, ISBN-1790308100.

Patrick H. Stakem, *Martian Railroad,* 2019, PRRB Publishing, ISBN-1794488243.

Patrick H. Stakem, *Exoplanets,* 2019, PRRB Publishing, ISBN-1731385056.

Patrick H. Stakem, *Exploiting the Moon,* 2019, PRRB Publishing, ISBN-1091057850.

Patrick H. Stakem, *RISC-V, an Open Source Solution for Space Flight Computers,* 2019, PRRB Publishing, ISBN-1796434388.

Patrick H. Stakem, *Arm in Space*, 2019, PRRB Publishing, ISBN-9781099789137.

Patrick H. Stakem, *Extraterrestrial Life*, 2019, PRRB Publishing, ISBN-978-1072072188.

Patrick H. Stakem, *Space Command*, 2019, PRRB Publishing, ISBN-978-1693005398.

CubeRovers, A Synergy of Technologys, 2020, PRRB Publishing, ISBN-979-8651773138.

Robotic Exploration of the Icy moons of the Gas Giants. 2020, PRRB Publishing, ISBN- 979-8621431006

Hacking Cubesats, 2020, PRRB Publishing, ISBN-979-8623458964.

History & Future of Cubesats, PRRB Publishing, ISBN-979-8649179386.

Hacking Cubesats, Cybersecurity in Space, 2020, PRRB Publishing, ISBN-979-8623458964.

Powerships, Powerbarges, Floating Wind Farms: electricity when and where you need it, 2021, PRRB Publishing, ISBN-979-8716199477.

Hospital Ships, Trains, and Aircraft, 2020, PRRB Publishing, ISBN-979-8642944349.

2020/2021 Releases

CubeRovers, a Synergy of Technologys, 2020, ISBN-979-8651773138

Exploration of Lunar & Martian Lava Tubes by Cube-X, ISBN-979-8621435325.

Robotic Exploration of the Icy moons of the Gas Giants, ISBN- 979-8621431006.

History & Future of Cubesats, ISBN-978-1986536356.

Robotic Exploration of the Icy Moons of the Ice Giants, by Swarms of Cubesats, ISBN-979-8621431006.

Swarm Robotics, ISBN-979-8534505948.

Introduction to Electric Power Systems, ISBN-979-8519208727.

Centros de Control: Operaciones en Satélites del Estándar CubeSat (Spanish Edition), 2021, ISBN-979-8510113068.

Exploration of Venus, 2022, ISBN-979-8484416110.

Patrick H. Stakem, *The Search for Extraterrestial Life,* 2019, PRRB Publishing, ISBN-1072072181.

The Artemis Missions, Return to the Moon, and on to Mars, 2021, ISBN-979-8490532361.

James Webb Space Telescope. A New Era in Astronomy, 2021, ISBN-979-8773857969.

www.ingramcontent.com/pod-product-compliance
Lightning Source LLC
Chambersburg PA
CBHW030514220526
45464CB00006B/2792